I0473635

Police Radar Operator Error

Microwave and Laser Radar Protocols and Results of Improper Use

Donald Sawicki

Police Radar Operator Error

© Copyright 2012 Donald S. Sawicki

All rights reserved.

Contents

Chapter 1 - Types of Police Radar

There are 2 basic types of police radar, **microwave** and **laser** (lidar) radar. Both measure speed, but use different methods that have some different limitations. Microwave police radar transmits a continuous (steady) microwave signal and measures the echo reflection Doppler Shift to determine speed. Lidar transmits infrared laser pulses to calculate range by measuring round trip pulse time, the change in range during one lidar cycle time measures speed.

Microwave and laser radars are designed to measure vehicle speed for vehicles traveling in a **straight direction on a flat** (level or incline) **surface**. If a vehicle's **speed is changing** greater than (plus or minus) ± **3 mph per second,** a radar/lidar cannot measure speed.

Microwave Radar

Microwave radars have relatively large beams and work well for close or long ranges (greater than **1000 - 3000 feet)** in light to moderate traffic. The wide beam allows easy aiming. A radar (antenna) can also be mounted to the patrol vehicle for **moving mode** or **stationary** operation. Microwave radars can be used from behind glass (windshield) inside the patrol vehicle.

Laser Radar (Lidar)

Laser radars have narrow beams that single out individual vehicles and works best for ranges less than about **500 feet** or so in light to dense traffic. The operator must select and **carefully aim** the lidar at a vehicle, typically the license plate. Lidar can only be operated from a **stationary** position (not a moving vehicle) and in the clear. Lidar should not point through glass such as the windshield or side windows.

1

Microwave versus Laser Radar

	Microwave	**Laser**
Operation	Stationary or Moving	Stationary
Aim (pointing)	Easy Aim or Fixed Mounted	**Exact** Aim Required
Traffic Conditions	Light to Moderate	Light to Dense
Range (distance)	Short to Long Range	Short Range
Measurements	Speed	Speed and Range
Transmissions	Continuous to seconds	Seconds

Some microwave radars can measure 2 vehicles at the same time, the strongest echo and the fastest.

Laser radars should **not** be operated from behind glass such as a windshield.

1.1 Microwave Radar

Microwave (and laser) police radars are designed to operate from as **close to the road** as possible. **The greater the distance from the traffic lane the radar is located the greater the limitations and chance for errors.**

Theory and Principles

The radar transmits a continuous microwave signal and simultaneously measures a vehicle's reflection echo that is frequency shifted (Doppler Principle) for moving vehicles. The *Doppler Principle* states the reflected frequency echo from a moving object is frequency shifted directly proportional to (vehicle) speed. The phenomenon is similar to a moving car horn or train whistle, approaching the pitch sounds high, even with the listener no change, moving away a lower pitch. The greater the speed the higher or lower the pitch.

Display / Indicator

In addition to displaying speed, most radars have a speaker for producing a **tone (pitch) proportional to the speed** of the target vehicle. The higher the tone pitch the greater the speed.

Modes

All radars can measure **on-coming** (closing) traffic from a stationary position, some can also measure **receding** (going) traffic, and some can measure on-coming and/or going traffic from a **moving** patrol vehicle. Some moving mode radars can also measure traffic in the **same-lane** (direction) as patrol vehicle.

Some radars can **measure 2 vehicles** at the same time, the strongest echo and the fastest echo while in stationary or moving mode.

3

Sample Time (Integration Period)

Typical **integration period**, *minimum time to get one measurement*, is about **0.3 seconds**, some radars require only 0.25 seconds to get one measurement. *It takes more than one integration period to establish and maintain a valid track*.

Frequency Bands

Three frequency bands are used in the USA for police radar, X band (1 channel), K band (1 channel), and Ka band (13 - 26 channels). Most Ka band radars operate on a single fixed (frequency) channel.

Police Microwave Radar Frequency Bands

Band	X	K	Ka
Frequency	10.525 GHz	24.150 GHz*	33.4 - 36 GHz
Wavelength	1.1 in 2.8 cm	0.49 in 1.2 cm	0.33 - 0.35 in 8.3 - 9.0 mm

* Some K band radars operate on 24.125 GHz

GHz -- gigahertz (1,000,000,000 Hertz) = 1 Billion Hertz
in -- inch
cm -- centimeter
mm -- millimeter

X band radars are used less because the spectrum is becoming crowded with other transmitters that cause interference. However X band has better **all weather** performance and typically a **longer detection range**.

K and **Ka** band radars have slightly better beams (more narrow than X band) and usually smaller antennas (smaller is easier to use/mount). **Weather** such as **rain** and fog degrade detection range.

Beamwidth

Beamwidth varies from model to model from about **9° to 25°**, the larger the antenna and/or the higher the frequency the more narrow the beam.

Typically Ka band photo radars (across the road radars) have a horizontal beamwidth of 5° and a vertical beamwidth 15° - 20°.

Beam Spread Equation

Beam Spread = 2 R *tan* (ß / 2)

R = Range,   ß = Beamwidth

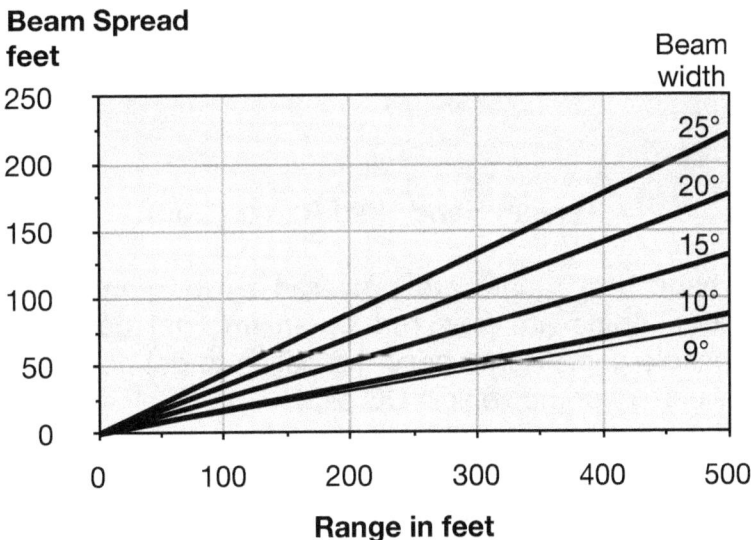

Range versus Beam Spread

Physical Configurations
Microwave police radars are available in 2 basic configurations, **hand held** or **fixed mounted** to patrol vehicle.

Hand Held Radar
Hand held radars usually operate from a **stationary position only** (not from a moving patrol vehicle) by design. Some hand held radars, by design, operate in a stationary or *moving mode*. These radars fit into a mount inside the patrol vehicle pointing the antenna out the windshield for moving mode operation.

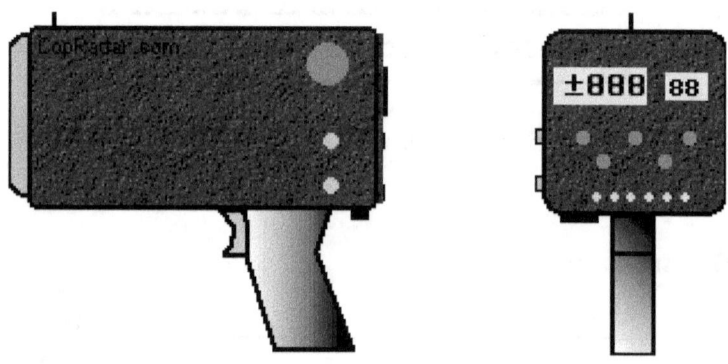

Typical Hand Held Radar (Gun)

Some hand held radars are **powered** by the **vehicle battery** (DC plug). Some are powered by an **internal battery** in the radar handle grip. Some have the option to get power from an internal battery or the vehicle DC plug.

Fixed Mounted Multi Unit Radar
Fixed mounted multi unit radars have a forward fixed antenna, some systems have an optional 2nd (rear) antenna. All fixed radars can operate in a **stationary** or **moving** mode.

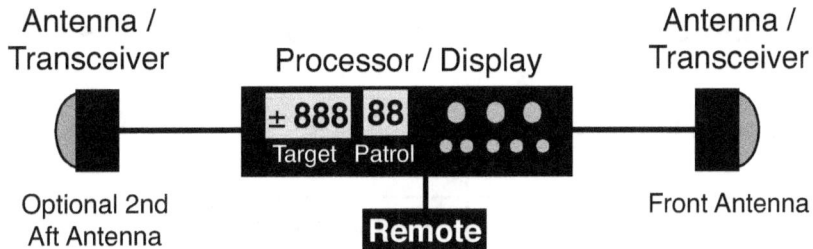

Typical Fixed Mounted Radar

The antenna for moving mode **must be pointed directly in front of patrol vehicle**, or *directly behind for aft antenna*, for accurate speed measurements. The patrol vehicle must also be aligned to traffic for stationary operation.

Some fix mounted systems have the transceiver in the Display / Processor unit instead of **antenna** housing. Antenna(s) can be mounted inside or outside patrol vehicle. Outside antennas must be hermetically sealed

Hand Held versus Fixed Mounted Radar

	Hand Held Radar	Fixed Mount Radar
Modes	Stationary	Stationary & Moving
Operation	Inside or Outside Patrol Vehicle	Patrol Vehicle must be aligned to traffic.

Moving Mode Radar

Moving mode radar requires and measures **patrol vehicle speed** to calculate traffic speed. Traffic echoes in moving more are the **closing speed** (patrol plus traffic speed). The radar subtracts patrol speed to get traffic speed.

Microwave Radar Specifications* / Options**

Configurations**	- Hand Held - Fixed Mounted
Frequency Bands**	- X Band; 10.525 GHz - K Band; 24.150 GHz - K Band; 24.125 GHz - Ka Band; 33.4 - 36 GHz
Operation**	- Stationary - Stationary / Moving

Modes**	- On coming Traffic
	- Receding Traffic
	- Strongest & Fastest Traffic
	- Moving Mode
	  -- opposite direction traffic
	  -- same-lane (direction) traffic

Operation**	3000 feet max (varies with target)
Beamwidth*	9° - 20°
Sample Time* Time to get 1 Measurement.	0.3 second typical
Accuracy*	± 1 mph stationary mode ± 2 mph moving mode ± 1 mph Patrol Speed moving mode
Indicators	Speed Display & Audio Tone
Power	Vehicle DC Plug &/or Internal

* Specifications may vary slightly with some models.

Common Worldwide Police Radar Frequencies

Band	Frequency	Wavelength	Notes
S	2.455 GHz	4.8 inches	Obsolete
X	9.410 GHz	1.25 inches	Europe
X	9.900 GHz	1.2 inches	Europe
X	**10.525 GHz**	**1.1 inches**	**USA**
Ku	13.450 GHz	0.88 inches	Europe
K	24.125 GHz	0.49 inches	Worldwide
K	24.150 GHz	0.49 inches	Worldwide
Ka	33.4-36.0 GHz	0.33-0.33 inches	Worldwide
IR	332 THz	904 nm	Worldwide Lidar

IR - Infrared
THz - Terahertz = 1 Trillion Hertz = 1,000,000,000,000 Hz
nm - nanometer = 0.000 000 001 meters

radar *(noun)* -

(1) acronym for **RA**dio **D**etection **A**nd **R**anging.
(2) a remote sensor that emits electromagnetic waves (radio, microwave, infrared laser light, etc.) in order to measure reflections for detection purposes (presence, location, motion, speed, etc.).
(3) radiolocation.
(4) field disturbance sensor.
(5) proximity sensor.

9

1.2 Photo Radar

Photo radars are **across the Road** radars that aim a narrow beam (typically 5° wide) across the road (typically **20°**), instead of down the road (close to 0°). The radar beam crosses only a very small segment of the road.

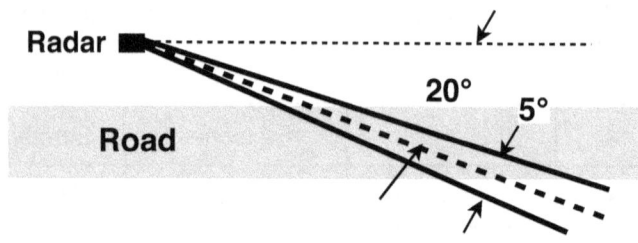

Typical Photo Radar Setup

A vehicle is in the beam only a fraction of a second. Additionally the vehicle Doppler reflection is distorted due to the alignment angle (20°) and vehicle wheels/tires rotating (close and directly in the beam). The alignment angle, distorted reflection and extra processing required makes across the road (photo radar) **inherently less accurate and less reliable** than conventional (down-the-road) radar.

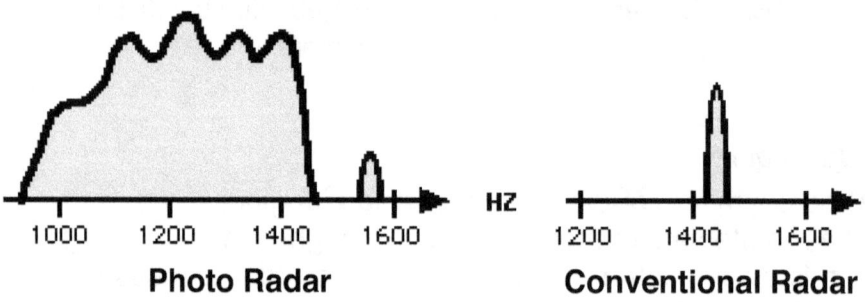

Photo Radar   **Conventional Radar**

Doppler Echo Spectrum for a 20 mph Vehicle
K Band Radar aligned 20°, 5° Horizontal Beamwidth

Spectrum (Doppler Shift) data from *Calibration Techniques for Across-the-Road Traffic Radars* -- National Institute of Standards and Technology, NIST Technical Note 1398, May 1998.

The radar beam must be **aligned properly,** to the design specification (typically 20°), for the radar to compute speed with *some* accuracy.   If the alignment **angle is shallow (smaller) calculated speed is HIGH**, if the angle is larger calculated speed is **low**.

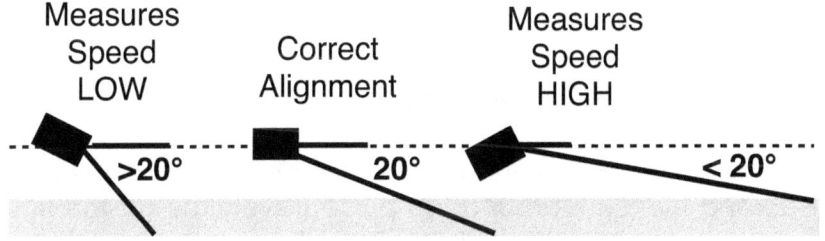

Photo Radar Correct and Incorrect Alignment

Speed Error % Based on Alignment & Align Error

$$\textbf{Speed Error \%} = 100 \left(\frac{cos(A \pm E)}{cos(A)} - 1 \right) \%$$

A = Alignment Design Angle (typically 20°).
E = ± Alignment Angle Error.

Design Alignment Angle A = 20°

Angle Error	Speed Error	Angle Error	Speed Error	Angle Error	Speed Error
1°	-0.7 %	-1°	0.6 %	-10°	4.8 %
2°	-1.3 %	-2°	1.2 %	-11°	5.1 %
3°	-2.0 %	-3°	1.8 %	-12°	5.4 %
4°	-2.8 %	-4°	2.3 %	-13°	5.6 %
5°	-3.6 %	-5°	2.8 %	-14°	5.8 %
6°	-4.4 %	-6°	3.3 %	-15°	6.0 %
7°	-5.2 %	-7°	3.7 %	-16°	6.2 %
8°	-6.0 %	-8°	4.1 %	-17°	6.3 %
9°	-6.9 %	-9°	4.5 %	-18°	6.4 %

1.3 Laser Radar (Lidar)

All police lidars are designed to operate from as **close to the traffic lane** as possible from a **stationary position**, not from a moving patrol vehicle. **The greater the distance from the traffic lane the lidar is located the greater the limitations and chance for errors.**

Theory and Principles

Lidar transmits narrow infrared (invisible to human eye) laser pulses and measures round trip pulse travel time (at the speed of light) from a reflective object. The change in range during 1 integrations period or sample time (typically 1/3 second) measures speed (change in range over 1/3 second). The lidar displays calculated speed and last range measurement.

Displays / Indicators

In addition to displaying **speed** and **range**, most lidars have a speaker for producing a **tone (pitch) proportional to the speed** of the target vehicle. The higher the tone pitch the greater the speed.

Beamwidth Divergence

Beamwidth varies with model from **3 - 4 milliradians** (mRad or mR), or 0.17° - 0.23°. Additionally the beam should be reflected from a **flat surface** and not overlap to any other part of the vehicle in front of or behind the *intended* surface. Lidar *User Manuals* typically recommend the operator aim at license plate / bumper / grill.

Beam Spread Equation

$$\text{Beam Spread} = 2\,R\,tan\,(ß\,/\,2)$$

R = Range,   ß = Beamwidth (divergence)

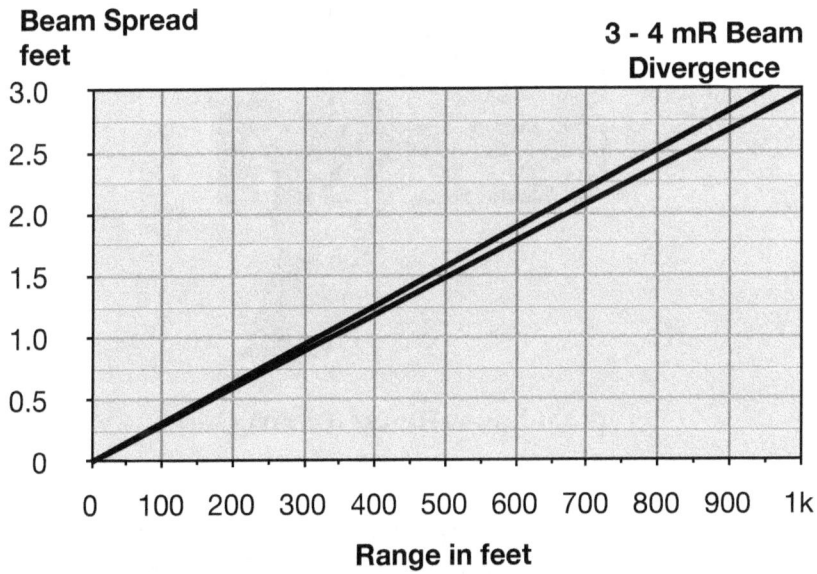

Beam Spread feet vs **Range in feet** — 3 - 4 mR Beam Divergence

Weather

Lidars are intended for use in **clear dry weather**. Atmospheric conditions can easily render a lidar useless by degrading detection range.

Conditions that Degrade Lidar

- **Fog**
- **Rain**
- Snow
- Sleet

- **Dust Particles**
- **Smoke**
- Carbon Dioxide (CO_2)

13

Configurations

Lidars are available in 2 basic configurations, **hand held** (gun) or **Binoculars style** (seldom used in USA). Both have aiming apparatus (lenses) for pin-pointing a vehicle.

Typical Laser Radar (Lidar) Gun

Some lidars use a single **aperture** for transmit and receive, some have separate transmit and receive apertures.

The **aim device** (lenses) enhances the view and superimposes an **aim Circle/Dot** or **crosshairs**, and **speed** and **range** so the operator can aim and see speed/range.

Typical **integration period**, minimum time to get one measurement, is about **1/3 second**. *It takes more than one integration period to establish and maintain a valid track*.

Some lidars are **powered** by the **vehicle battery** (DC plug). Some are powered by an **internal battery** in the lidar handle grip. Some have the option to get power from an internal battery or the vehicle DC plug.

Infrared Pulsed Laser Specifications

Configurations	- Hand Held - Binoculars
Transmit Wavelength	904 nm ± 5 nm Infrared (IR)
Operation	Stationary
Useful Range	500 feet
Beamwidth	3 - 4 milliradians 0.17° - 0.23°
Modes	- On-coming Traffic - Receding Traffic
Integration Period Time to get 1 Measurement.	1/3 second typical
Accuracy	± 1 mph Speed ± 1 foot Range (typical)
Indicators	Speed Display/ Audio Tone Range Display
Power	- Vehicle DC Plug - Internal Battery - Vehicle or internal Battery

nm - nanometer = 0.000 001 meters

DEFINITIONS
Laser - **L**ight **A**mplification by **S**timulated **E**missions of **R**adiation.
Lidar - **LI**ght **D**etection **A**nd **R**anging
Ladar - **LA**ser **D**etection **A**nd **R**anging

Chapter 2 - Radar Test and Checks

2.1 Calibration

Microwave and laser radars are precision remote sensing instruments and should be tested throughly by a qualified laboratory **at least once or twice per year**. Testing requirements vary with locality. Police should track (log) all calibration tests, many require a *Calibration* sticker be attached to the radar.

CALIBRATION	
BY: _____	ID: _____
DATE: _____	DUE: _____
RADAR: _____	MODEL: _____
S/N: _____	PATROL CAR: _____

Calibration testing should include;
- Test Entire Speed Range
- Range Accuracy (lidar only)
- Transmit Frequency or Wavelength
- Transmitter Power
- Radiated Power
- Beam Divergence (Beamwidth)
- Receiver / Detector Sensitivity
- Power Supply Limits (± volts)
- Tuning Fork Calibration (microwave Radar)

Most microwave radars come equipped with at least one tuning fork for testing speed. *Moving mode radar requires 2 tuning forks* (different resonances) to test moving mode, one fork simulates a target vehicle, the other simulates patrol vehicle speed.

Tuning Forks

A radar can measure a vibrating tuning fork and produce a speed reading proportional to the **resonant frequency** (tone) of the fork (fork resonance equals *Radar Doppler* shift). The displayed speed is a function of radar **transmit frequency** and tuning fork resonance frequency. The tuning fork is not traveling at the speed the radar measures (radar is tricked).

Speed Reading given Resonance
All Radar Bands (including Ka)

$$V = 0.3353 \, (\, f_r / f_x \,)$$

V = Speed Reading in **mph**
f_x = Radar Transmit Frequency in **GHz**
f_r = Tuning Fork Resonance in Hertz (**Hz**)

Speed Reading given Resonance
X and K Band Radar

	X Band 10.525 GHz	K Band 24.125 GHz	K Band 24.150 GHz
Speed (mph)	$0.03186 \, f_r$	$0.01390 \, f_r$	$0.01388 \, f_r$

f_r = Tuning Fork Resonance in Hertz (**Hz**)

Tuning forks should be used daily to quick check radar speed accuracy. To check a microwave radar with a tuning fork the radar must be transmitting and the fork must be vibrating. One of the sides (not front/rear) of the tuning fork should be placed within a few inches of the antenna.

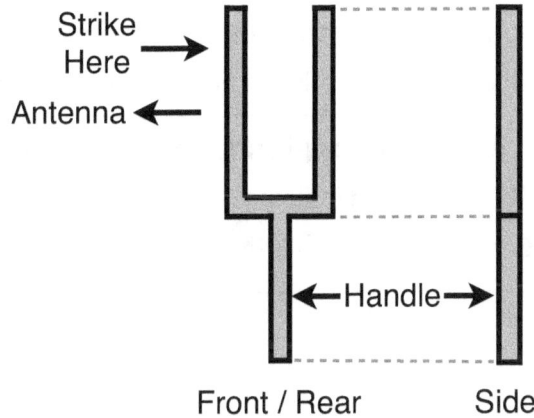

Typical Tuning Fork

Tuning forks should be calibrated (tested and logged)
every time the radar is calibrated.

CALIBRATION		
BY: _____	ID: _____	
DATE: _____	DUE: _____	
RADAR FREQ: _____ Ka 34.6 ____GHZ		
TUNING FORK S/N: _____		
FORK RESONANCE: _____ 4160 ____ HERTZ		
SPEED: _____ 40.3 ___ MPH		

SUMMARY
• Radar / Lidars should be in **calibration** period.   - Logs / Sticker should indicate next calibration date.
• Microwave radar **tuning fork(s)** should be calibrated.   - Logs / Sticker should indicate next calibration date.   - **Moving mode** requires 2 forks, different speeds.

2.2 Operator Testing

After ensuring the radar or lidar is **in calibration**, and the tuning fork(s) in calibration (microwave radar) the operator should *run built-in self test*. Most radars or lidars run self-test automatically on power up. *Self-test should also be initiated during and at end of use (officer duty shift).*

MICROWAVE RADAR TUNING FORK TEST

Microwave radars should be tested with a **tuning fork**. The radar must be transmitting as the operator places a calibrated vibrating tuning fork within a few inches of the antenna. The radar should read a speed that is proportional to the tuning fork resonance (tone).

The tuning fork, at a minimum, should be labeled with the speed the radar should read and the tuning fork resonance tone. For example a *K band* radar (*24.150 GHz*) should read *55 mph* for a tuning fork that resonates anywhere between *3930 -3990 Hertz*.

MOVING MODE TUNING FORK TEST

Moving mode radar requires **2 tuning forks** with different tones vibrating at the same time. The radar must be transmitting and in moving mode. The **patrol speed** should read the **lower pitch** tuning fork, **target vehicle** speed should read the **difference** between both tuning fork speeds.

Moving Mode Tuning Fork Test Example
K Band Radar (**24.150 GHz**)

	1st Tuning Fork	2nd Tuning Fork
Resonance	2880 Hz	1800 Hz
Speed	**40 mph**	**25 mph**

Displayed Speeds

Target Vehicle	Patrol Vehicle
15 mph	**25 mph**

Displayed Target Speed is difference between fork speeds (40 - 25 = 15).

MICROWAVE RADAR INTERFERENCE TEST

Most radars have a **Receive-Only mode** (transmitter off) for checking interference. Operator should scan antenna in different directions checking for *false readings* or interference detected by the radar RFI (*Radio Frequency Interference*) circuits.

LASER RADAR TEST

Range accuracy is critical for lidar to calculate an accurate speed. An object of known distance should be measured in *Range only mode*, and measure within specifications (typically ± 1 foot).

Beam / Sight alignment is critical for lidar to single out vehicles. A utility pole or stop sign located 200 - 500 feet distance are good test targets for alignment (**and range**) test. The laser is swept pass the test target, a range reading should only occur when the test target is in the reticle or aim circle area. To check vertical alignment the laser should be held at 90° (right or left 90 degrees).

RADAR AND LASER RADAR

Test Vehicle

Many police departments require the radar or lidar be tested using a **controlled test vehicle** with a **calibrated speedometer**. Several different speed runs should be conducted. Ideally the test runs should be conducted at the location the radar/lidar will be used.

SUMMARY
• Check Radar / Lidar in **Calibration**. -- Check tuning fork(s) calibrated (microwave radar).
• Run Radar / Lidar **Self-Test**. -- Should run at start, during, and end of shift.
Microwave Radar • **Tuning Fork Test.** -- Moving mode requires 2 tuning forks. • **Test for Interference** *(receive only mode).*
Laser Radar (Lidar) • **Test Check Range.** • **Test Check Alignment.**
• Test using Controlled **Test Target Vehicle**. -- Test vehicle should have calibrated speedometer.

Chapter 3 - Operational Setup

3.1 Setup Location
Microwave and Laser Radar

Microwave and laser radar should only be used to measure traffic speed on a **straight and flat** (level or incline) road and traveling at a **constant speed**. The radar or lidar should be position as **close to traffic lane(s)** as possible. Too far off the road imposes a blind region, traffic too close to measure.

Proper Setup

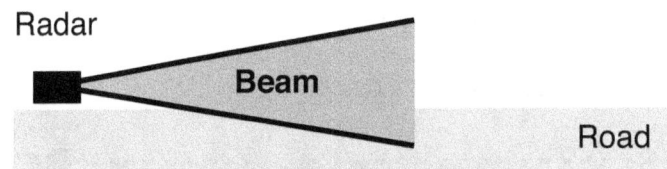

- Radar/Lidar should be **close to road** to *minimize Cosine Effect Errors*.
- Road must be **straight and flat** (level or incline).
- Traffic must be at a **constant speed**.

Worse Possible Setup

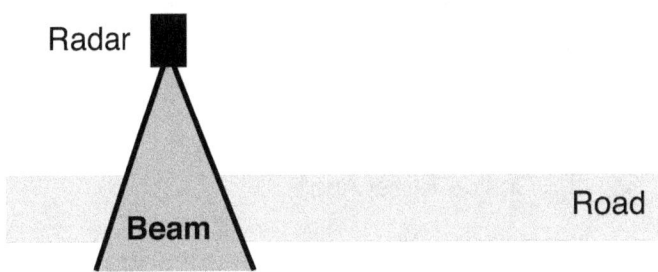

In the *worse possible setup* above the beam (microwave or laser) is pointed in the direction the traffic speed, relative to the radar/lidar, is 0 mph. It does not matter how far or close the radar/lidar is to the road in this setup, speed will measure 0 mph. This is a result of the *Cosine Effect* explained below.

Cosine Effect Error

Microwave and laser radar measures speed relative to the radar. Traffic traveling directly toward (0°) the radar is measured exactly. A few degrees off angle and a slight error occurs. The greater off angle the greater the error. At 90° speed relative to the radar is 0 mph. This phenomenon is called the **Cosine Effect** because the error is proportional to the cosine of the vehicle angle **off 0°** (directly at radar).

Cosine Effect Angle

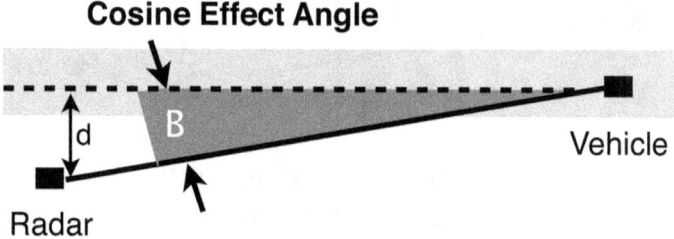

The angle the antenna / aperture is pointing does not have any effect on the Cosine Error, only the **angle to the target vehicle**.

Measured Speed Due to the Cosine Effect

$$V_m = V_o \; cos \; (B)$$

V_m = Radar Measured Speed
V_o = Target Vehicle True Speed
B = Cosine Effect Angle

The Cosine Effect results in speeds measuring low, the closer the vehicle the lower the speed.

3.2 Blind Zone (Too Close)
Microwave and Laser Radar

The *Cosine Effect* results in speeds measuring low and imposes a **minimum range** or blind zone. When the cosine angle changes too fast, relative speed changes too fast to measure.

The **closer** a moving vehicle the **faster** the Cosine Effect angle changes, and the faster the Doppler echo changes. When the Doppler echo is changing faster than radar/lidar accuracy, the echo signal cannot be tracked because (Doppler) frequency (caused by the Cosine Effect) is changing too fast.

The blind zone is a function of **radar / lidar distance from the vehicle lane** and **vehicle speed**. The greater off the road and the faster the traffic, the larger the blind zone.

Minimum Range (Blind Zone)

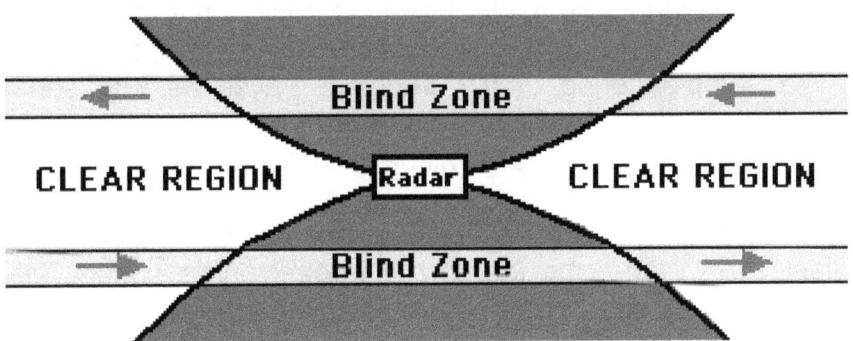

Minimum Range Equation

$$R_{min} = \sqrt{\frac{(V_o\, d)^{4/3}}{(V_{acc}\, /\, t_i)^{2/3}} - d^2} + \frac{V_o}{t_i}$$

R_{min} = Minimum Range (Distance to end of Blind Zone)
V_o = True Vehicle Speed
d = Radar/Lidar Distance from Vehicle Lane
V_{acc} = Radar/Lidar Accuracy (typically **1 mph**)
t_i = Radar/Lidar Integration Period (typically **1/3 second**)

Minimum Range Equation from *Police Traffic SPEED RADAR Handbook*, p 35-41, ISBN 9781456524289, 2011.

The square root term is the range the cosine effect acceleration is equal to radar accuracy, the last term is distance traveled for radar to complete one full cycle (one measurement).

Blind Zone Table

The following table list minimum range based on a radar or lidar with accuracy of ± **1 mph accuracy** (±1.47 feet per second), and a sample period of **0.3 seconds**. Systems with shorter sample periods will have slightly shorter (smaller) blind zones. Systems with longer periods will have slightly larger blind zones.

The table far left column is distance radar/lidar is from the target vehicle traffic lane, the top row is target vehicle speed. Table data is **blind zone range** in feet, traffic closer is too close to measure.

For example a **radar/lidar located 35 feet** from center of vehicle lane *cannot* measure traffic traveling at **60 mph** at ranges closer than **146 feet** (almost 50 yards).

Blind Zone Range in Feet '

From Lane feet '	35 mph	40 mph	45 mph	50 mph	55 mph	60 mph	65 mph	70 mph
0'	15'	18'	20'	22'	24'	26'	29'	31'
5'	39'	43'	47'	52'	56'	60'	64'	68'
10'	52'	58'	63'	69'	74'	80'	85'	90'
15'	63'	70'	76'	83'	89'	96'	102'	108'
20'	72'	80'	88'	95'	103'	110'	117'	124'
25'	80'	89'	98'	107'	115'	123'	131'	138'
30'	88'	98'	108'	117'	126'	135'	143'	152'
35'	95'	106'	117'	127'	137'	146'	155'	165'
40'	102'	114'	125'	136'	146'	157'	167'	177'
45'	108'	121'	133'	145'	156'	167'	178'	188'
50'	114'	127'	140'	153'	165'	177'	188'	199'
55'	119'	134'	148'	161'	173'	186'	198'	209'
60'	125'	140'	155'	168'	182'	195'	207'	219'
65'	130'	146'	161'	176'	190'	203'	216'	229'
70'	135'	152'	167'	183'	197'	212'	225'	239'
75'	139'	157'	174'	190'	205'	220'	234'	248'
80'	144'	162'	180'	196'	212'	227'	242'	257'

SUMMARY
• Locate Radar / Lidar as **Close to Road** as Possible.
• Know **Blind Zone Limitations**.

Chapter 4 - Moving Mode Use

4.1 Moving Mode Considerations
Microwave Radar

Moving mode radar should only be used to measure traffic speed on a **straight and flat** (level or incline) road and traveling at a **constant speed**. Patrol vehicle must also be at a constant speed. As with stationary radar, the greater the distance between patrol and target vehicle lanes, the more significant the *Cosine Effect*.

Moving radar **tracks 2 reflections**, a *target vehicle* reflection and the *ground reflection*. The target vehicle echo is Doppler shifted by the closing (or opening) speed between radar and target (*sum of patrol speed plus target speed*). The ground reflection (large signal echo) is Doppler shifted by patrol vehicle speed. The radar subtracts out the patrol speed from closing or opening speed to get target speed.

Antenna Alignment in Moving Mode
The **ground echo** used to measure patrol speed must be reflected from **directly in front** of the patrol vehicle (*directly aft for rear antennas*).   The more off angle (either side) the reflection the lower the patrol speed measurement (Cosine Effect). **A low patrol speed measurement causes a high target speed calculation**. Target vehicle speed calculates high by exactly the speed the patrol reads low.

Patrol Speed Shadowing
Any off angle reflected object can and will cause moving radar to measure patrol speed low. This phenomenon doesn't happen all the time, but is common enough to have a name - *patrol speed shadowing*.   A number of objects are known to cause this, including other vehicles whether moving or not.

Objects that cause Patrol Speed Shadowing

• concrete sound barriers • concrete construction barriers • metal guardrails • cable crash barriers • overhead and roadside signs
• bridge trusses • hillcuts / depressions / tunnels • underpass / pillars
• highway barrels (metal/water filled plastic) • construction zones (signs and equipment)
• plowed snow pile • ice patches • ditch water
• parked vehicles • moving vehicles near patrol vehicle

Patrol Speed Shadowing induces a speed error for the patrol speed measurement and the resulting calculated vehicle speed. The error can be a few mph or 10's of mph, and may be monetary or last an extended period.

To manage patrol speed shadowing some moving radars monitor patrol vehicle speed using the vehicle's VSS (Velocity Speed Sensor -- Speedometer). If the radar measured patrol speed does not match closely the VSS reading, speed measurements are not displayed (until speeds match again). Some radars automatically switch to stationary mode when patrol vehicle is not moving. Side note: Vehicle speedometers are not accurate enough for radar speed calculations.

4.2 Blind Zone Moving Radar
Microwave Radar

Moving mode radar measures the closing, or opening, speed between patrol and target vehicles (sum of both speeds). This increases the blind zone by increased speed (patrol plus target). This does not apply to same-lane radar that has different restrictions

Moving Mode Blind Zone Range in Feet '
Speed is CLOSING or OPENING

Between Lanes feet '	70 mph	80 mph	90 mph	100 mph	110 mph	120 mph	130 mph	140 mph
0'	31'	35'	40'	44'	48'	53'	57'	62'
5'	68'	76'	84'	92'	99'	107'	114'	121'
10'	90'	100'	110'	119'	129'	138'	147'	156'
15'	108'	120'	131'	143'	154'	164'	175'	185'
20'	124'	137'	150'	163'	175'	188'	200'	211'
25'	138'	153'	168'	182'	196'	209'	222'	235'
30'	152'	168'	184'	199'	214'	229'	243'	257'
35'	165'	182'	199'	216'	232'	248'	263'	278'
40'	177'	195'	214'	231'	249'	265'	282'	298'
45'	188'	208'	228'	246'	265'	283'	300'	317'
50'	199'	220'	241'	261'	280'	299'	317'	335'
55'	209'	232'	254'	275'	295'	315'	334'	353'
60'	219'	243'	266'	288'	309'	330'	350'	370'

Calculations for above table use ± **1 mph accuracy** (± 1.47 feet per second) and **integration time of 0.3 seconds**.

Accuracy for target vehicle is typically ± 2 mph, patrol accuracy ± 1 mph. The patrol accuracy is used to calculate blind zone for moving mode radar (not target speed accuracy).

A moving **radar 35 feet from target lane** with closing speed of 100 mph (patrol speed plus target speed) *cannot* measure traffic at ranges closer than **216 feet** (72 yards). The exact speeds of the patrol vehicle or target vehicle do not matter, only closing (or opening) speed.

Same-Lane (Direction) Mode

Same-lane (direction) mode does not have a minimum range restriction, but it does have some speed restrictions. The **patrol vehicle** must be traveling at least **5 - 20 mph**. The *target vehicle* must be traveling at *least 10 mph*, and speed must be greater than or less than the patrol speed by at least ± **5 mph.**

Chapter 5 - Interference

Microwave radars should only be used in areas relatively **free of external signal and electronic noise** sources, **transmitters** and **power lines** are the most common.

Interference will be detected automatically by the **RFI** (Radio Frequency Interference) circuits in many instances, but not all. Many radars have a *receive only mode*, receiver on but not transmitting. This allows an operator to check for some types of interference the RFI circuits miss. The radar should not get any speed readings in *receive only mode*, if it does there is an interference problem.

Interference Effects
• **Reduces Detection Range** automatic gain control adjust out interference and detection range.
• **Activates Interference Detection Circuits (RFI)** Some models stop processing speeds if interference detected.
• **Mask Legitimate Targets** Unbeknownst to the operator.
• **Produce False Speed Readings** Unbeknownst to the operator.

5.1 Patrol Vehicle Transmitters

Patrol vehicle transmitters can/do cause interference to the radar by over-the-air transmissions or though the vehicle DC power lines. When operating radar these devices should be in standby mode, not transmitting (off would be better).

Patrol Vehicle Transmitters
• 100 watt VHF/UHF Radio
• 2 watt VHF/UHF Walkie-Talkies
• High Frequency (HF) Radio
• Citizen's Band Radio (HF)
• Cellular Phone (UHF)
• Digital Communications Systems
• Satellite Uplink (microwave bands)

A 100 watt radio, a 2 watt walkie-talkie, and the citizens band (CB) radio caused interference in most or all 7 radars tested by the *National Highway Traffic Safety Administration**.

* *Police Traffic Radar ISSUE PAPER*, U.S. Department of Transportation, National Highway and Safety Administration , February 1980, DOT HS-805 254

5.2 Field Disturbance Sensors

Field disturbance sensors are transceivers used for short range detection of objects and/or motion. Numerous sensors operate *exactly* on police radar frequencies (**X and K bands**) and can be detected by and/or interfere with police radar.

A radar may or may not detect this type of interference because these signals look exactly like a legitimate radar echo. A **radar in receive-only mode** can detect these signals. **Radar detectors** easily detect these signals (as a radar signal).

Common Field Disturbance Sensors

Application	Use
Intrusion Sensor Burglar Alarms	Commercial Stores Warehouses Public Buildings
Automatic Door Opener	Commercial / Grocery Stores Vehicle entrances Railroad entrances Airports
Obstruction Detectors	Fork Lifts Moving Farm Equipment Railroad Locomotives Railroad Rolling Stock
Object Detectors	Factory Production Lines
Speed Measuring	Train Yards Race Tracks Sporting Events Car / Motorcycle Clubs

Field disturbance senors radiate less power then a police radar, but it is *much larger* than an echo from a vehicle.

5.3 Other Interference and Effects

Radio Frequency Interference (RFI)

Interference occurs when unwanted radio frequency energy gets into the radar circuits. Most interference gets into the radar antenna / receiver, strong energy sources can get into almost any/all circuits.

High power transmitters generate noise well outside the transmitting band and adds to other background noise. Additionally these transmitters can generate harmonics and intermodulation products all over the frequency spectrum.

High Power Signal and Noise Sources

- **Airports**
  Radars (weather, tracking)
  Communications
  Beacons
- **Communication Towers**
  TV / AM / FM Broadcast
  Microwave / Radio Relay Towers
  Fire / Police / Medical Services
  Cellular Services
- **Military Installations**
- **Telephone / Cable Antenna Farms**
- **Amateur Radio**
- **Satellite Uplinks**
- **Weather Radars**
- **Aircraft / Marine Vessels**

Mechanical Interference

A phenomenon well known among many radar operators is that pointing the radar antenna at the police car heater/air conditioning fan (while rotating) can produce false readings.

Objects that can produce readings include but not limited to;

- Vehicle Heater/AC Fan
- Moving Signs (highway billboards)
- Building / House Air Conditioning Fans
- Wind Gauges

Natural Interference

Rain, **fog**, and **humidity** effect K and Ka band radar detection range, X band to a lesser extent. A *moderate rain can render a radar useless*, drizzle and fog can severely degrade detection range.

Temperature affects the sensitivity of all receivers: the lower the temperature, the better the sensitivity (longer range). The sun affects background noise; as the sun heats objects, the thermal excitation of atoms in conducting materials generates noise that can obscure signals and reduce a receiver's performance. Lightning can also interfere with a receiver's performance.

UNINTENTIONAL RADIATION

Devices that are high voltage or draw a lot of electrical current, especially fast changing voltages or currents, can and do interfere with radar and radio receivers.

Outdoor Lighting

Fluorescent, mercury and sodium vapor lighting, especially switching ON or OFF, generates broadband noise that causes interference. Many **neon / argon signs** intentionally switch ON/ OFF to attract attention, more interference.

Power Lines and Equipment

Power generating / transmission equipment produces noise interference. This type of interference usually causes a buzzing or humming of the radar audio Doppler and/or false speed readings.

Vehicle Electrical Noise

Patrol **vehicles**, or other near-by vehicles generate noise. Ignitions, alternators, spark plugs and wiper motors, **especially if faulty**, can interfere with radar.

Lighting Noise Sources
• Fluorescent Lights • Mercury Lights • Sodium Vapor Lights • Neon / Argon Signs
AC Power Noise Sources
• Power Plants • Generator Sub-Stations • Transformers • Transmission Lines
Vehicle(s) Noise Sources
• Automotive Ignitions • Alternators • Spark Plugs • Wiper Motors

Chapter 6 - Microwave Radar Errors

6.1 Stationary and Moving Radar

Vehicle in Blind Zone

All radars have a blind zone, traffic too close to radar. The blind zone results from radar timing (minimum time to get a measurement), and the *Cosine Effect* angle. The blind zone can be **10's of feet to 100's** or more depending on radar distance off vehicle lane and traffic speed. Also see Chapter 3.2 - Blind Zone and Chapter 4.2 - Blind Zone Moving Radar.

Microwave radars are designed to measure the strongest echo reflection, usually the closest. A vehicle in the **blind zone** may be the closest, but NOT the closest that can be measured. This can lead to target *vehicle misidentification*. The radar is not measuring the closest vehicle, it is measuring the closest vehicle in the clear region.

Vehicle Radar Size (reflection)

Larger vehicles tend to have a larger radar reflection than smaller vehicles. A small closer vehicle can be mistaken (**misidentified**) for a larger more distant vehicle by the operator. Reflection strength depends on the range of the close small vehicle, and the difference between vehicle reflectivity strengths.

Vehicle reflectivity is measured by **Radar Cross Section (RCS)** and has dimensions of area (usually **square meters**). RCS is not the same as surface area even through the unit dimensions are the same. RCS varies with frequency, different frequencies produce different RCS sizes.

41

Measured Radar Cross Sections (RCS) in Square Meters

	X Band		Ka Band	
	Front	Rear	Front	Rear
1970 Dodge Pickup	200	300	84	670
1972 Oldsmobile Cutlass	100	10	110	73
1972 Dodge Dart	100	100	290	110
1970 AMC Gremlin	200	32	84	73
1965 Ford Mustang (convertible)	10	100	55	42
1966 Chevrolet Corvette	10	32	28	28
1967 Ford Cortina	10	10	84	42
Bicycle Coasting	20		2	
Bicycle Pedaling		2		7
Man Walking	1	1	3	4

Source: System Considerations for the Design of Radar Braking Sensors, IEEE Transactions on Vehicular Technology, vol VT-26, no. 2, page 151-160, May 1977.

RCS Approximations in Square Meters

Motor cycle	Small Car	Mid Size	Large Car	Pickup	RV
15	30	60	120	200	400

RV - Recreational Vehicle

Equation for Equal Signal Reflections

$$R_2 = R_1 \sqrt[4]{\dfrac{RCS_2}{RCS_1}}$$

R_1 = Range of Small Close Vehicle
RCS_1 = Radar Cross Section of Small Close Vehicle
R_2 = Range of Distant Large Vehicle
RCS_2 = Radar Cross Section of Distant Large Vehicle

The following tables illustrate distances for different size vehicles with **equal strength reflections**. All vehicles in the same row have equal reflections at the ranges listed. *Any vehicle that is closer will have a larger reflection.*

Ranges Vehicles have Equal Reflections
Approximate RCS values in square meters (sq m)

Motorcycle	Small Car	Mid Size	Large Car	Pickup	RV
RCS = 15 sq m	30	60	120	200	400
100 ft	119 ft	141 ft	168 ft	191 ft	227 ft
200 ft	238 ft	283 ft	336 ft	382 ft	454 ft
300 ft	357 ft	424 ft	505 ft	573 ft	682 ft
400 ft	476 ft	566 ft	673 ft	764 ft	909 ft
500 ft	595 ft	707 ft	841 ft	955 ft	1136 ft

X Band (10 GHz) Radar
Ranges Vehicles have Equal Echoes

Vehicle Front

65' Mustang 67' Cortina 66' Corvette	72' Cutless 72' Dart	70' Pickup 70' Gremlin
RCS = 10 sq m	100	200
100 ft	178 ft	211 ft
200 ft	356 ft	423 ft
300 ft	533 ft	634 ft
400 ft	711 ft	846 ft
500 ft	889 ft	1057 ft

Vehicle Rear

67' Cortina 72' Cutlass	70' Gremlin 66' Corvett	65' Mustang 72' Dart	70' Pickup
RCS = 10 sq m	32	100	300
100 ft	134 ft	178 ft	234 ft
200 ft	267 ft	356 ft	468 ft
300 ft	401 ft	533 ft	702 ft
400 ft	535 ft	711 ft	936 ft
500 ft	669 ft	889 ft	1170 ft

Ka Band (35 GHz) Radar
Ranges Vehicles have Equal Echoes

Vehicle Front

66' Corvett	65' Mustang	70' Pickup 70' Gremlin 67' Cortina	72' Cutless	72' Dart
RCS = 28 sq m	55	84	110	290
100 ft	118 ft	132 ft	141 ft	179 ft
200 ft	237 ft	263 ft	282 ft	359 ft
300 ft	355 ft	395 ft	422 ft	538 ft
400 ft	474 ft	526 ft	563 ft	718 ft
500 ft	592 ft	658 ft	704 ft	897 ft

Vehicle Rear

66 Corvett	65' Mustang 67' Cortina	70' Gremlin 72' Cutless	72' Dart	70' Pickup
RCS = 28 sq m	42	73	110	670
100 ft	111 ft	127 ft	141 ft	221 ft
200 ft	221 ft	254 ft	282 ft	442 ft
300 ft	332 ft	381 ft	422 ft	664 ft
400 ft	443 ft	508 ft	563 ft	885 ft
500 ft	553 ft	635 ft	704 ft	1106 ft

Police radar tracks largest (some can also track fastest) signal reflection, not always the closest.

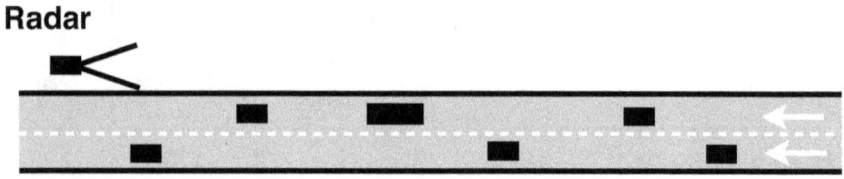

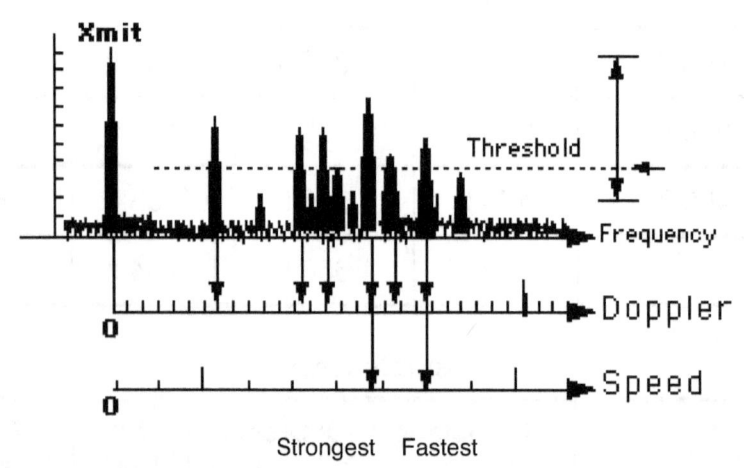

Strongest   Fastest

Doppler Radar View of the World

6.2 Moving Radar

Moving mode radar has additional operational considerations, **antenna alignment** and **patrol speed shadowing**. Radar measures patrol speed to determine traffic speed, any patrol speed errors translate directly into traffic speed errors.

Antenna Alignment

The antenna must be **aligned directly to the front** (0°) of the patrol vehicle to receive **ground reflections** from directly in front (*aligned to back for rear antennas*). Off angle reflections will cause the patrol speed to measure low (*Cosine Effect*) introducing errors into the target speed calculation. The more off angle (off 0°), the greater the speed error.

Measured Patrol Speed

$$V_m = V_p \, cos \, (\, B \,)$$

V_m = Measured Patrol Speed
V_p = Actual Patrol Speed
B = Angle Off 0°

Patrol Speed Error Percentage
for Off Angle Ground Reflection

Off Angle	Error	Off Angle	Error
0°	0%	50°	36%
10°	2%	60°	50%
20°	6%	70°	66%
30°	13%	80°	83%
40°	23%	90°	100%

With moving mode radar a low patrol speed reading causes a **high target speed reading** by the **amount patrol speed is low**. *Same-lane* mode has one exception, when *target vehicle slower* than patrol vehicle, target speed calculates low.

Patrol Speed Shadowing

Patrol speed shadowing forces the ground reflection to angle off, common in certain locations.   See Chapter 4.1 - *Moving Mode Considerations* for objects known to cause patrol speed shadowing.

Patrol Speed Shadowing from Moving Vehicles

Other moving vehicles **traveling slower and in front** (behind for aft antennas) of the patrol vehicle can also produce patrol speed shadowing. The radar picks up the other vehicle echo as the ground echo resulting in an **incorrect patrol speed** reading. Measured patrol speed will be the difference between the patrol vehicle speed and slower vehicle speed.

Moving Vehicle Patrol Speed Shadowing
Shadowing Vehicle Slower than Patrol Vehicle.

Measured Patrol Speed
Patrol Measured = Patrol Actual Speed - Shadowing Vehicle Speed

Opposite Direction Traffic Speed
Traffic Measured = Traffic Actual Speed + Shadowing Vehicle Speed

Same-Lane Traffic Speed
Traffic Measured = Traffic Actual Speed - Shadowing Vehicle Speed

EXAMPLE
- Radar traveling 35 mph uses 20 mph vehicle as ground echo.
- Measured patrol speed is **15 mph** (35-20).
- Opposite direction traffic measures high by 20 mph.
- Same direction traffic measures low by 20 mph.

Measure Traffic Speed Due
to Moving Vehicle Shadowing

Shadowing Vehicle Slower than Patrol Vehicle

Traffic Mode	Calculated Traffic Speed	Speed Error
Opposite Direction	$V + V_s$	High by Shadowing Vehicle Speed (V_s)
Same-Lane Traffic	$V - V_s$	Low by Shadowing Vehicle Speed (V_s)

V = Traffic True Speed
V_s = Shadowing Vehicle Speed (False Ground Echo).

Measured Patrol Speed = V_p - V_s

V_p = Patrol Vehicle Speed

Note; same direction vehicles traveling faster than patrol vehicle will not be mistaken for a ground reflection because the Doppler shift is negative.

49

Batching or Speed Bumping

Batching (Speed Bumping) occurs when the patrol speed suddenly changes speed causing a false and/or inaccurate speeding reading. Radar-measured traffic speed and patrol vehicle speed are not updated simultaneously; if patrol vehicle speed changes suddenly the radar may still be using outdated patrol data leading to an error. Sudden acceleration can cause batching or speed bumping.

Situations that cause
Batching (Speed Bumping)

- Acceleration (hit the gas)
- Braking
- Turning (curves)
- Hitting bumps

Chapter 7 - Laser Radar (Lidar) Errors

7.1 Potential Problems

Vehicle in Blind Zone

All lidars have a blind zone, traffic too close. The blind zone results from lidar timing (minimum time to get a measurement), and the *Cosine Effect* angle. The blind zone can be **10's of feet to 100's** or more depending on lidar distance off vehicle lane and traffic speed. Also see Chapter 3.2 - Blind Zone.

Interrupted Signal

If for any reason signal returns (echoes) are interrupted a speed / range error is possible. A vehicle passing between the intended target and lidar, or the lidar beam striking a tree, branches/leafs, sign, utility pole or tower, will cause problems.

Beam Alignment

Aim device and beam must be precisely in alignment. Some lidars have a separate transmit and receive aperture. Both apertures must be precisely aligned to each other and the aim device. **Alignment should be tested before use**.

Aiming

Lidars must be aimed (a steady aim) with care at a flat surface on the target vehicle. The greater the distance off the road (and the closer the vehicle), the more movement required to track the same spot. While lidars are hand-held, it is easier to track a moving vehicle surface point with the unit mounted to a tripod for a steady accurate aim.

Scanning Error

Scanning or sweeping the lidar beam across the ground such that range increases (or decreases) in a steady manner will produce a speed reading. The measured speed is a function of the change in range over time (beam scan or sweep time). Increasing range (short to long range) produces a receding speed, decreasing range (long to short range) produces an approaching speed reading.

For example a lidar will register a speed of 85 mph when the beam scans the ground (or guardrail etc.) approximately 125 feet per second (20 to 145 feet in one second).

Speed for Scan Rate Range

Speed	Scan Rate feet / second	Speed	Scan Rate feet / second
20 mph	29 ft/s	55 mph	81 ft/s
25 mph	37 ft/s	60 mph	88 ft/s
30 mph	44 ft/s	65 mph	95 ft/s
35 mph	51 ft/s	70 mph	103 ft/s
40 mph	59 ft/s	75 mph	110 ft/s
45 mph	66 ft/s	80 mph	117 ft/s
50 mph	73 ft/s	85 mph	125 ft/s

Scan Rate in feet/second = (22/15) x MPH

Scan Rate in feet/second = 1.47 x MPH

7.2 Speed / Range Errors

SPEED ERROR DUE TO **AIM ERROR**
A lidar beam that *moves* or *drifts* from the *back of the vehicle toward the front* **measures speed high** (measured **range is long**). If the beam moves from *the front toward the back* **measured speed is low** (measured change in **range is short**).

SPEED ERROR DUE TO **LONG RANGES**
Range errors, and thus speed errors, can occur inadvertently at ranges greater than about 500 feet. At 500 feet the beam covers a width of about 1.5 feet. This is wide enough to cover different parts of the vehicle (different ranges)   At longer ranges the beam will cover most or all of the vehicle increasing range error possibilities.

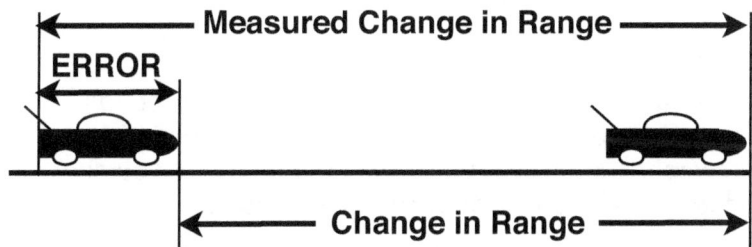

Measured Change in Range LONG
Speed Calculates HIGH

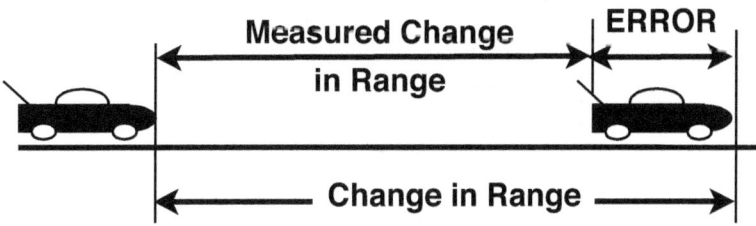

Measured Change in Range SHORT
Speed Calculates LOW

Speed Error

$$V_{err} = \pm\, d_{err}\, /\, (n\, t_i)$$

V_{err} = Speed Error
d_{err} = Distance Error
t_i = Lidar Sample Time (time to get 1 speed measurement)
n = a whole number (number of lidar samples)
$(n\, t_i)$ = Track Time (number of samples multiplied by sample time)

Using feet and seconds to get Error in mph

$$V_{err} = (15/22)\, (\pm\, d_{err})\, /\, (\, n\, t_i\,)$$
$$V_{err} = (0.682)\, (\pm\, d_{err})\, /\, (\, n\, t_i\,)$$

| V_m in **mph** | d in **feet** |
| ti in **seconds** | d_{err} in **feet** |

EXAMPLE
- Lidar with a 1/3 second sample time (3 samples/second).
- Measures vehicle speed in 1 second (3 sample periods).
- Lidar sweeps from passenger compartment to front bumper introducing a + 5 foot error during that 1 second.

Speed Error = (15/22) x 5 / (3 x 1/3) = +3.4 mph

Measuring Time = 1 second

Distance Error	Speed Error
5 feet	3.4 mph
10 feet	6.8 mph
15 feet	10.2 mph
20 feet	13.6 mph

Chapter 8 - Before and After a Ticket

8.1 Motoring Tips

If one knows of or is warned (from a CB or a passing motorist flashing headlights) of an **impending speed trap**, it's a good idea to drop speed 5 mph below the posted limit. This may or may not prevent a ticket, but could help in court if a ticket is issued. If one can claim the vehicle speed was at least 5 mph below the limit, speedometer accuracy will not be as questionable. It's not a bad idea to have the speedometer calibrated by a qualified shop, even after a ticket, so one may determine if the speedometer was accurate and if not the error.

If a driver believes (such as a warning from a radar detector or a visual sighting) a radar has just started illuminating (tracking) the vehicle, a good countermeasure is to gently **decelerate** at a rate of about **3 or 4 mph per second**. A traffic radar/lidar will experience problems measuring vehicles decelerating (or accelerating).

DUMMY RADAR

Some roads have unattended fixed mounted Dummy Radars that constantly transmit a signal in one of the radar bands, usually **K** or **Ka**. The FCC requires unattended transmitters to radiate less power than police radar. These transmitters (Dummy Radar) do not have receivers and cannot, by design, measure speed or anything else.

Dummy radars are *intended to set off radar detectors* to alert (fool) drivers (with detectors) to travel at the speed limit. Locals usually figure out where the dummies are and ignore them, but be careful because **police might be operating (like to operate) radar** on a different frequency (than the dummy) in the area. A popular place to mount dummy radar is on overhead highway signs, or portable message road signs.

Microwave Detectors

Microwave radar detectors can be helpful but be careful. Some police switch on the radar only when ready to target a vehicle. Depending on the situation a driver may only have a fraction of a second to respond, and in some cases may not be warned at all until after the radar gets a speed reading. The lesson: do not depend solely on a radar detector to warn a traffic radar is in the area.

Microwave radar detectors should be **located as high** as possible and **clear of any obstructions** (except windshield). The higher (above ground) the detector antenna the better the coverage (receive beamwidth).

Laser Radar (Lidar) Detectors

Laser radars (lidars) have a narrow beam making it difficult to detect unless the beam is **aimed directly at a laser detector**, or the beam is wide enough (range greater than 500 feet) to illuminate the detector. Laser radar detectors are generally not nearly as effected as microwave detectors.

Laser detectors stand the best chance to catch a lidar signal when the aperture (antenna lens) is mounted near the license **plate, bumper or grill**. At close range the lidar pulses may still miss the detector. Detectors inside the vehicle passenger compartment have little chance to detect a signal at close range. The windshield also reduces the detector's *sensitivity*, decreasing detection.

Traffic Stop Tips

If the police light you up (pull over) stop at the *closest safest* place for you and the officer, at night try to find a well lighted area. Curves and blind spots are poor stop locations. Once stopped most officers prefer you stay in your vehicle. When approached be polite and keep your hands visible -- at night turn on your interior doom light before the officer approaches.

Typically the officer request to see a driver license (and/or proof of insurance and/or vehicle registration), and ask if you know why you were stopped. Comply and answer honestly. Do not try to explain away why you were speeding with an excuse (most are lame), it almost never works and is an admission of guilt.

Common excuses* for speeding (DO NOT USE)
• It was unintentional.
• I was in a hurry (appointment, pick up someone, etc.).
• I was being forced to speed (by a tailgater).
• I think the limit is too low.
• My modern car stops more quickly than cars on the road at the time the limit was set.
• I don't think the same limit should apply when the road is empty late at night.
• The limit does not apply to me because I am an above average driver.
• My speeding's acceptable because it's not much over the limit and others abuse the limit more flagrantly.

* Factors influencing driver speed choices, Department of Transport (England), Circular Roads 1/95.

Many citations require a bond (driver's license or cash) and signature. Before signing a traffic citation READ IT. Make sure you understand the violation(s).

8.2 The Courtroom

CAVEAT: This document is not intended to give legal advice in any way. This section and any other sections or references to the law or courtroom activities is based solely on the author's personal experiences and observations.

Several factors must be considered before deciding to fight a speeding charge. First and most important, you must be wrongly accursed -- do not try to beat a charge if you're guilty. However, if you were not speeding as fast as accused, you might plead for the lesser speed violation to lower the fine (if you can convince the judge). With some insurance companies the higher the speed the more the insurance rate increases on conviction. The court clerk should know if the fine would change for a lower speed, and your insurance agent should know if rates change with speed.

The **Manual on Uniform Traffic Control Devices** (MUTCD), established by federal law in 1966, sets national standards for every **sign**, signal, pavement marking, and traffic control signal device in the USA. The MUTCD requires speed limits (*USE of speed limit signs*) are to be determined by an **Engineering Study** (as defined by MUTCD 1A.13). An Engineering Study must also be done (and documented) before any speed limit can be changed. If some government body changes a speed limit **without a proper study** -- *the speed limit is illegal*.

Preparation for Court

Be prepared to document as many **facts** as possible. Ground photographs and/or aerial photographs of the site could be helpful. Show approximate distances from the map scale or actual measurements. Note approximate locations for vehicle and radar at start and finish of track. Be careful any maps or photographs are not outdated.

DISCOVERY

Some experts suggest a defendant should file a Discovery motion or a request for Discovery form with the court clerk for some or all of the following items.

Discovery Motion Items
FCC Public Safety Radio Services (or Radar) License
Radar documents -- Make, Model, Serial Number, Options, Age -- Manufacturer Certificate of Calibration -- Operator Manual and Specifications -- Calibration Log Sheets (date, **due**, lab)
Tuning Fork documents (microwave radar) -- Specifications - Band, Speed, Resonance, Tolerance) -- Calibration Log Sheets (date, **due**, lab).
Officer Training / Qualifications -- Certificate of Competency -- Training Material -- Officer Training Records
All Policies Pertaining to Radar Use
Officer Notes and log book

FCC license and Tuning Forks are not applicable to laser radar (lidar).

OFFICER PRESENTATION

Most officers are trained to collect certain information in the event the case is disputed. Below list typical minimum information an officer should be prepared to present in court.

Police Officer Presentation
Establish **time**, **place**, **location** of radar.
Establish **location** of **offending vehicle**.
Identify (able to) the offending vehicle.
Establish and identify the **vehicle operator**.
Visually observed apparent excessive speed.
Observed vehicle was alone out front. (not fastest mode)
Established **steady stable track** history.
Insured **minimum interference** sources.
Establish Radar **tested** before (and/or after) use; -- self-test, -- tested with tuning fork((not laser radar), -- tested for Range Accuracy (*laser radar*), -- Scope / Beam alignment tested (*laser radar*), -- tested against test vehicle.
State and document **qualifications** and **training**.

YOUR DEFENSE

Microwave and laser radars are accurate speed measuring devices -- *when properly used*. Improper use makes speed measurements unreliable and can introduce errors resulting in an unjust speeding citation.

Operators must follow setup and operational procedures for accurate results. Not following proper protocol introduces *reasonable doubt*.

Make Sure Officer Followed Proper Test Protocol

Before Use
• Check Radar / Lidar in **Calibration**.
• Run Radar / Lidar **Self-Test**. -- Should run at start, during, and end of shift.
• Test using Controlled **Test Target Vehicle**. -- Test vehicle should have calibrated speedometer. Some agencies do not require a controlled test vehicle.
Microwave Radar • Check tuning **fork(s) calibrated** • **Tuning Fork Test** -- Moving mode requires 2 tuning forks. • **Interference Check** -- radar in *receive only mode* - *Moving Mode Radar* • Insure Antenna(s) **Aimed Directly Forward** (or Aft)
Laser Radar (Lidar) • **Test Check Range**. • **Test Check Alignment**.

Make Sure Officer
Followed Operational Procedures

Operational Use
• Road must be **Straight and Flat** (lever or incline)
• Radar / Lidar **Close to Traffic Lanes**
• Traffic **Speed Constant** (± 3 mph / sec)
• Know **Blind Zone Range** (too close to measure)
• Insure no **Interference** Sources (microwave mainly)
• Run Radar / Lidar **Self-Test**. -- Should run at start, during, and end of shift.
Microwave Radar • Make sure **Traffic Spacing Adequate** -- Ignore Readings for Multiple Closely Spaced Vehicles.
Moving Mode • Watched for **Patrol Speed Shadowing** -- none observed and no sources in area.
Laser Radar (Lidar) • Beam should be **Aimed at Flat Surface** on Vehicle • Ranges should be **less than 500 feet** to minimize Speed Errors Due to Range Errors

Cosine Effect Defenses

LARGE COSINE EFFECT ANGLE

A common occurrence is the radar cosine effect angle is so large that when measured speed is corrected for the cosine effect angle corrected speed is excessively high and unlikely. This would indicate the radar was tracking another signal such as another vehicle echo or interference from a transmitter.

Speed Corrected for Cosine Effect

$$V_o = V_m / cos(B) = V_m \ sec(B)$$

sec = secant angle

V_m = Radar Measured Speed
V_o = Target Vehicle True Speed
B = Cosine Effect Angle

Cosine Angle	Corrected Speed	Percent
10°	1.02 V_m	+ 2 %
20°	1.06 V_m	+ 6 %
30°	1.15 V_m	+ 15 %
40°	1.31 V_m	+ 31 %
50°	1.56 V_m	+ 56 %
60°	2.00 V_m	+100 %
70°	2.92 V_m	+ 192 %

For example if the cosine angle is 60° corrected speed is double measured speed.

63

If the officer stated observed estimated vehicle speed to be about the same as the radar measured speed, one could argue the officer's judgment was biased incorrectly by the radar. If the officer's estimated speed conflicts with radar corrected speed, which one is correct, if any?

OBSERVED COSINE EFFECT
The cosine effect causes measured speed to change, the closer the target vehicle the lower the measured speed to actual. A good track (steady stable reflection) will show cosine effects as the vehicle approaches the blind zone. A 75 mph vehicle approaching a radar (microwave or lidar) will display speed readings that sequence something like below.

75 ...   74-74-74-74-74-73-73-72- *Blind Zone*

Each "-" is a sample period. The speed readings are predictable and should be noticeable to the officer (especially when the vehicle is relatively close to the radar) -- if not the radar was not tracking the intended (expected) vehicle.

Officer Visual Speed *Estimate*
Some courts (judges) will accept an officer's meager visual observation of vehicle speed (even if all other evidence discarded). This is dubious at best. Any court accepting a visual speed guess alone as good enough is a kangaroo court by definition.

Test the officer's ability to estimate the speed of a falling object. Drop an object from several different heights and ask the officer to guess the speed of the object as it hits the ground. Note that a falling object has a changing speed, speed increases at a constant rate (acceleration due to gravity). Someone (the officer) adapt at visually judging speed should be expected to estimate a falling object's speed (instantaneous) upon striking the ground, or at least the average speed -- half (0.5) instantaneous speed.

Free Falling Object

Distance	Time	Speed
2 feet	0.35 sec	7.7 mph
3 feet	0.43 sec	9.5 mph
4 feet	0.50 sec	10.9 mph
5 feet	0.56 sec	12.2 mph
6 feet	0.61 sec	13.4 mph
7 feet	0.66 sec	14.5 mph
8 feet	0.71 sec	15.5 mph
9 feet	0.75 sec	16.4 mph

Free Falling Object Speed and Time Equations

$$V = a\,t \qquad\qquad t = \sqrt{\frac{2\,d}{a}}$$

V = speed
d = distance object falls
t = time object falls start to stop
a = acceleration due to gravity (32.2 feet / sec*sec)

Police 10 Codes

Police and other agencies use **"10" codes** to verbally communicate efficiently. The word 10 indicates the next number (or numbers) is code. Four codes are widely used throughout the United States and listed below.

APCO - Association of Police Communications Officers

Code	General Purpose	APCO	Norfolk, VA	Walnut Creek, CA
10-0		Use Caution		
10-1	Unable to Copy - Change Location	Signal Weak	Police Officer Needs Help	Poor Radio Reception
10-2	Signal Good	Signal Good	Assist Officer	Stop Transmission
10-3	Stop Transmitting	Stop Transmitting	Clear the Air - Emergency	Stop Transmission
10-4	Acknowledgment (OK)	**Affirmative (OK)**	Acknowledgment (OK)	**Message Received**
10-5	Relay	Relay To / From	See a Complainant	Relay Message
10-6	Busy- Unless Urgent	Busy	Investigation Police or Fire	Change Radio Channel
10-7	**Out of Service**	**Out of Service**	**Out of Service - Off Air Subject to Call**	**Out of Service**
10-8	**In Service**	**In Service**	**In Service**	**In Service**
10-9	Repeat	Say Again	Arrive at Scene	Repeat Message
10-10	Fight in Progress	Negative	Traffic Detail	Off Duty
10-11	Dog Case	_On Duty (Employee Number)	Broken Glass	Visitors Can Hear Radio
10-12	Stand By (Stop)	Stand By (Stop)	Vandalism	Advise Weather / Road Conditions

Police 10 Codes

Code	General Purpose	APCO	Norfolk, VA	Walnut Creek, CA
10-13	Weather- Road Report	Weather Conditions	Leaking Water Main or Sewer Hole in Street / Sidewalk	
10-14	Prowler Report	Message / Information	Convoy or Escort	
10-15	Civil Disturbance	Message Delivered	Prisoner in Custody	Prisoner in Custody
10-16	Domestic Problem	Reply to Message	Pick Up Prisoner	Pick Up
10-17	Meet Complainant	En-route	Administrative Assistance	Getting Fuel
10-18	Quickly	Urgent	Detail	
10-19	Return to ___	(In) Contact	Return to Station	Return or Go to ___
10-20	**Location**	**Location**	**What is Your Location**	**Location**
10-21	**Call (_) by Phone**	**Call (_) by Phone**	**Call (_) by Phone**	**Telephone**
10-22	Disregard	Disregard	Investigate a Break-In	Cancel or Disregard
10-23	Arrived at Scene	Arrived at Scene	Breaking-In (In Progress)	Stand-By
10-24	Assignment Completed	Assignment Completed	Someone in the Building	
10-25	Report in Person (Meet)	Report To (Meet)	Prowler	Do You Have Contact With ___?
10-26	Detaining Subject, Expedite	Estimated Arrival Time (ETA)	Larceny	Clear of Warrants
10-27	(Driver) License Information	License / Permit Information	Rape Report	Subject Wanted
10-28	**Vehicle Registration Information**	**Vehicle Information**	**Check FUll Registration, License, Motor, Name, Stolen**	**Registration Check**
10-29	Check for Wanted	Records Check	Person with a Gun	Check for Warrants

Police 10 Codes

Code	General Purpose	APCO	Norfolk, VA	Walnut Creek, CA
10-30	Unnecessary Use of Radio	Danger / Caution	(a) Vehicle Accident (b) Vehicle Accident Personal Injury (c) Hit and Run	
10-31	Crime in Progress	Pice Up	Hold Up and Robbery	
10-32	Man with Gun	__ Units Needed (Specify)	Defective Traffic Light	
10-33	Emergency	Need Immediate Assistance	Execute Warrant	Alarm is Sounding
10-34	Riot	Current Time	Narcotics Investigation	
10-35	Major Crime Alert		Get a Stolen Auto Report	Time Check
10-36	**Correct Time**		**Correct Time**	**Correct Time**
10-37	(Investigate) Suspicious Vehicle		Finished with Last Assignment	Please Identify Your Unit
10-38	Stopping Suspicious Vehicle		(a) Reckless Driving (b) Drunk Driving	
10-39	Urgent-Use Light, Siren		Report of a Dead Person	Can __ Come to Radio?
10-40	Silent Run-No Light, Siren	Fight in Progress	Suspicious Person-Auto	Is __ Available for Phone Call?
10-41	Beginning Tour of Duty	Beginning Tour of Duty	(a) Lost Child (b) Investigate Runaway	
10-42	Ending Tour of Duty	Ending Tour of Duty	Car Improperly Parked	
10-43	Information	In Pursuit	Drunk	
10-44	Permission to Leave __ for __	Riot	Disturbance (type)	
10-45	Animal Carcass at __	Bomb Threat	FIght	Subject Condition: A to D

Police 10 Codes

Code	General Purpose	APCO	Norfolk, VA	Walnut Creek, CA
10-46	Assist Motorist	Bank Alarm	Attempt Suicide	
10-47	Emergency Road Repair at __	Complete Assignment Quickly	(a) Injured (b) Sick (c) Demented Person (d) Maternity Case, State, Civilian, Fire, or Police	
10-48	Traffic Standard Repair at __	Detaining Suspect, Expedite	Person Overboard	
10-49	Traffic Light Out at __	Drag Racing	Braking Dog	Proceeding to __
10-50	Accident (F,PI,PD)	Vehicle Accident, PD,PI, F	Court Cases	Drugged
10-51	Wrecker Needed	Dispatch Wrecker	General Message	Drunk
10-52	Ambulance Needed	Dispatch Ambulance	Open Door/ Window (State Which)	Ambulance Needed
10-53	Road Blocked at __	Road Blocked	Gas-Repairs-Wash	Person Down
10-54	Livestock on Highway	Hit and Run Accident, PD, PI, F	Man Molesting Children	Possible Body
10-55	Intoxicated Driver	Intoxicated Driver	Bomb Threat	Coroner's Case
10-56	Intoxicated Pedestrian	Intoxicated Pedestrian	Unruly Crowd	Suicide (a) Attempted
10-57	Hit and Run (F, PI, PD)	Request RT Operator	Tampering With Automobile	
10-58	Direct Traffic	Direct Traffic	Burglar Alarm	
10-59	Convoy or Escort	Escort	Traffic Violator	Security Check

Police 10 Codes

Code	General Purpose	APCO	Norfolk, VA	Walnut Creek, CA
10-60	Squad in Vicinity	Suspicious Vehicle	(a) Dead Dog (b) Live DOg (c) Female or Stray (d) Dog Bite	
10-61	Personnel in Area	Stopping Suspicious Vehicle	Void IBM Card	Bike Theft
10-62	Reply to Message	B and E in Progress	Radio Test	
10-63	Prepare Make Written Copy	Prepare to Receive Assignment	Personal Relief	Prepare to Copy
10-64	Message for Local Delivery	Crime in Progress	Eating (State Location)	
10-65	Net Message Assignment	Armed Robbery	Exposure	
10-66	Message Cancellation	Notify Medical Examiner	Send Wrecker to (a) Owner Request (b) Police Request	Suspicious Person
10-67	Clear for Net Message	Report of Death	Smoke & Flames Visible	Person Calling for Help
10-68	Dispatch Information	Livestock in Roadway	In Commission on Stand-By	
10-69	Message Received	Advise Telephone Number	Held Up By Bridge or Train	
10-70	Fire Alarm	Improper Parked Vehicle	Anger / Caution	Prowler
10-71	Advise Nature of Fire	Improper Use of Radio	False Alarm	Shots Fired
10-72	Report progress on Fire	Prisoner in Custody	Person Found in Burning Building	
10-73	Smoke Report	Mental Subject	Existing Conditions	How Do You Copy?
10-74	Negative	Prison / Jail Break	En-route	
10-75	In Contact with __	Wanted or Stolen	Dispatch Mechanic	

Police 10 Codes

Code	General Purpose	APCO	Norfolk, VA	Walnut Creek, CA
10-76	En Route __	Prowler	Rewind Box (Give Location)	
10-77	ETA (Estimated Time of Arrival)	Direct Traffic at Fire Scene	Send VEPCO (State Gas or Electric) (b) Send C and P	
10-78	Need Assistance		Held Up by (state)	
10-79	Notify Coroner		Courtesy Call	
10-80	Chase in Progress	Fire Alarm	Critical Call (Code Red)	Explosion
10-81	Breatherlizer Report	Nature of Fire	Alarm of Fire	
10-82	Reserve Lodging	FIre in Progress	Additional Engine Co.	
10-83	Work School Crossing at __	Smoke Visible	Additional Ladder Co.	
10-84	If Meeting __ Advise ETA	No Smoke Visible	Second Alarm	
10-85	Delay Due to __	Respond without Blue Lights / Siren	Third Alarm	
10-86	Officer / Operator on Duty		Person Trapped	Any Traffic for Me?
10-87	Pickup / Distribute Checks		Auto Fire	
10-88	Present Telephone # of __		Request Deputy Chief	Provide Cover for Units
10-89	Bomb Threat		Request Additional Chief	
10-90	Bank Alarm at __		Transfer Fire Alarm Wire	
10-91	Pick Up Prisoner / Subject		Check Fire Alarm Box or Master Box	Hazard
10-92	Improperly Parked Vehicle		Fire Alarm Circuit Open or Trouble on Circuit	

Police 10 Codes

Code	General Purpose	APCO	Norfolk, VA	Walnut Creek, CA
10-93	Blockade		Fire Alarm	
10-94	Drag Racing		Request Gas or Diesel Fuel	
10-95	Prisoner / Subject in Custody		Grass or Trash Fire	
10-96	Mental Subject		In Quarters	
10-97	Check (Test) Signal		Signal Weak	Arrived at Scene
10-98	Prison / Jail Break		Signal Good	Completed Assignment
10-99	Wanted / Stolen Indicated		Fireman Need Help	
10-101	What is Status? (Are you secure?)			
10-106	Secure (Status is secure)			

Speed of Light / Sound

Speed of Light in a Vacuum

299,792,456.2	meters / second
983,571,050.5	feet / second
1,079,252,842	kilometers / hour
670 616 625.4	miles / hour
582,749,914.9	knots / hour
299,792.4562	kilometers / second
186,282.3959	miles / second
161,874.9763	knots / second
983.571 050 5	feet / microsecond
0.983 571 051	feet / nanosecond

microsecond = 0.000 001 seconds
nanosecond = 0.000 000 001 seconds
Light travels about 1 foot in 1 nanosecond

Approximate Speed of Sound (Mach 1)

Sea Level		36,000 - 82,000 feet		82,000 - 154,000 feet
765 mph 1 mi / 4.7 sec	Linear Decrease to -->	660 mph 1 mi / 5.5 sec	Linear Increase to -->	755 mph

Speed Conversions

1 meter / second =	1 / 0.3048	feet / second
	9 / 2.5	kilometer / hour
	1 / 0.44704	miles / hour
	900 / 463	knots
1 foot / second =	0.3048	meters / second
	1.09728	kilometers / hour
	15 / 22	miles / hour
	274.32 / 463	knots
1 kilometer / hour =	2.9 / 9	meters / second
	5 / 5.4864	feet / second
	1 / 1.609344	miles / hour
	1 / 1.852	knots
1 mile / hour =	0.44704	meters / second
	22 / 15	feet / second
	1.609344	kilometer / hour
	402.336 / 463	knots
1 knot =	463 / /900	meters / second
	463 / 274.32	feet / second
	1.852	kilometer / hour
	463 / 402.336	miles / hour

knot = nautical mile per hour

Distance Conversions

1 inch =	0.0254	meters
1 foot =	0.3048	meters
1 yard =	0.9144	meters
1 meter =	1 / 0.0254	inches
	1 / 0.3048	feet
	1 / 0.9144	yard
1 kilometer =	1 / 1.852	nautical miles
	1 / 1.609344	miles
	1 / 0.0009144	yards
	1 / 0.0003048	feet
1 mile =	1609.344/1852	nautical miles
	1.609344	kilometer
	1760	yards
	5280	feet
	1609.344	meters
1 nautical mile =	1852/1609.344	miles
	1.852	kilometer
	1852 / 0.9144	yards
	1852 / 0.3048	feet
	1852	meters

1 nautical mile (U.K.) = 1853.184 meters

Constants

Typical US Interstate

Lane Width:	12 feet
Left (inside) Shoulder Width Rural or Urban: 3 or more lanes each direction: Heavy Truck Traffic: Mountains: Mountains ,4 or more lanes each direction:	 4 feet 10 feet 12 feet 4 feet 8 feet
Right (outside) Shoulder Width Rural or Urban: Heavy Truck Traffic: Mountains:	 10 feet 12 feet 8 feet
Median Width Rural: Mountains or urban:	 36 feet 10 feet
Vertical Clearance Rural: Urban: Sign supports /pedestrian overpasses:	 16 feet 14 feet 17 feet

Acceleration Due to Gravity

32.1740   feet / sec sec
9.80665   meters / sec sec

Standard Railroad Gauge (US)
4 feet, 8.5 inches

Index

www.ingramcontent.com/pod-product-compliance
Lightning Source LLC
Chambersburg PA
CBHW071609170526
45166CB00003B/1036